Ticks

by Barbara Keevil Parker

Lerner Publications Company • Minneapolis

To Maggie, Mitzi, Chocolate, Kaila, Hannah, Trigger, and Tess

My appreciation goes to Harry Lerner, who suggested I write a book on ticks. Thanks to my researcher, Duane F. Parker; to Linda Crotta Brennan for reviewing and commenting on this manuscript; and to my editor, Wendy Aalgaard.

A special thanks to Dr. Roger A. LeBrun, Carnegie Professor of Life Sciences, University of Rhode Island, for reviewing the text.

The photographs in this book are used with the permission of: Scott Bauer/Agricultural Research Service, USDA, pp. 4, 36, 38 (left); © Eye of Science/Photo Researchers, Inc., p. 5; © James H. Robinson/Photo Researchers, Inc., p. 6; © Joe McDonald/Visuals Unlimited, p. 7 (left); © Leroy Simon/Visuals Unlimited, p. 7 (right); © Jim Kalisch, Department of Entomology, University of Nebraska-Lincoln, p. 8; © Getty Images, pp. 9, 35, 43; Centers For Disease Control Public Health Image Library (CDC)/Jim Gathany, pp. 10, 12, 20, 47, 48 (bottom); © Dr. Robert Calentine/Visuals Unlimited, p. 11; © Dr. James L. Castner/Visuals Unlimited, pp. 13, 16, 22, 29, 30, 31, 32, 33, 37 (right), 46; © Dwight R. Kuhn, p. 14; © Dr. Ken Greer/Visuals Unlimited, p. 15; © Karlene Schwartz, pp. 17, 18, 38 (right), 40, 48 (top); © Todd Strand/Independent Picture Service, p. 19; © Brad Mogen/Visuals Unlimited, pp. 21, 27; © David H. Ellis/Visuals Unlimited, p. 23; © Darwin Dale/Photo Researchers, Inc., p. 24; © Noah Poritz/Photo Researchers, Inc., p. 25; © Medical-on-line/Alamy, p. 26; CDC/J. S. Wiseman, Texas State Health Department/Dr. Pratt, p. 28; © Gary Meszaros/Visuals Unlimited, p. 34; © Steve Maslowski/Visuals Unlimited, p. 37 (left); © Larry Mulvehill/Photo Researchers, Inc., p. 39; © Hank Morgan/Photo Researchers, Inc., p. 41; © Wegner/ARCO/naturepl.com, p. 42. Front cover: CDC/Jim Gathany.

Text copyright © 2007 by Barbara Keevil Parker

Lerner Publications Company
A division of Lerner Publishing Group
241 First Avenue North
Minneapolis, MN 55401 U.S.A.

Website address: www.lernerbooks.com

Library of Congress Cataloging-in-Publication Data

Parker, Barbara Keevil.
    Ticks / by Barbara Keevil Parker.
        p.   cm. —   (Early bird nature books)
        Includes index.
    ISBN-13: 978–0–8225–6464–5 (lib. bdg. : alk. paper)
    ISBN-10: 0–8225–6464–5 (lib. bdg. : alk. paper)
    1. Ticks—Juvenile literature. I. Title. II. Series.
    QL458.P436 2007
    595.4'29—dc22   2006018810

Manufactured in the United States of America
1 2 3 4 5 6 – JR – 12 11 10 09 08 07

Contents

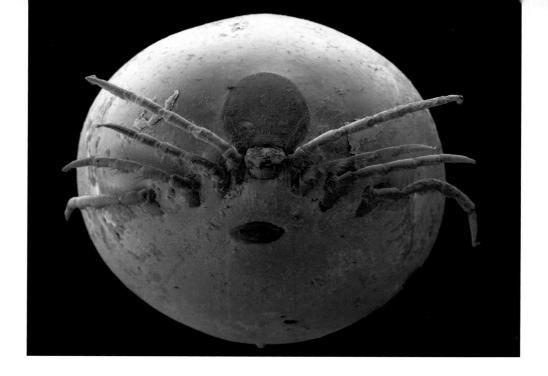

Be a Word Detective

Can you find these words as you read about ticks?
Be a detective and try to figure out what they mean.
You can turn to the glossary on page 46 for help.

arachnids   **larvas**   **parasites**
capitulum   **Lyme disease**   **questing**
host   **molting**   **scutum**
hypostome   **nymphs**

There is more than one way to form plurals of some words.
The word larva *has two possible plural endings—either an*
e or an s. *In this book,* s *is used when many larvas are*
being discussed.

Chapter 1

Both of these creatures are ticks. What kind of animals are ticks?

Meet a Tick

Nick and his dog Tess play all day. They romp in the woods next to Nick's yard. At dinnertime, Nick's mother notices a little brown bug. It is crawling on Nick's pants.

A few days later, Nick finds a funny bump behind Tess's ear. The bump is almost as big as an M&M. It is soft and stuck to Tess's head. Nick and his mom found ticks.

Ticks look like bugs. But they are not insects. Insects have only six legs. Adult ticks have eight legs. Ticks are arachnids (uh-RACK-nihdz). Spiders, daddy longlegs, scorpions, and mites are also arachnids. Arachnids have eight legs but no antennas and no wings.

Scorpions (left) *and spiders* (right) *are relatives of ticks.*

About 850 different species, or kinds, of ticks live in the world. Around 80 tick species are found in the United States.

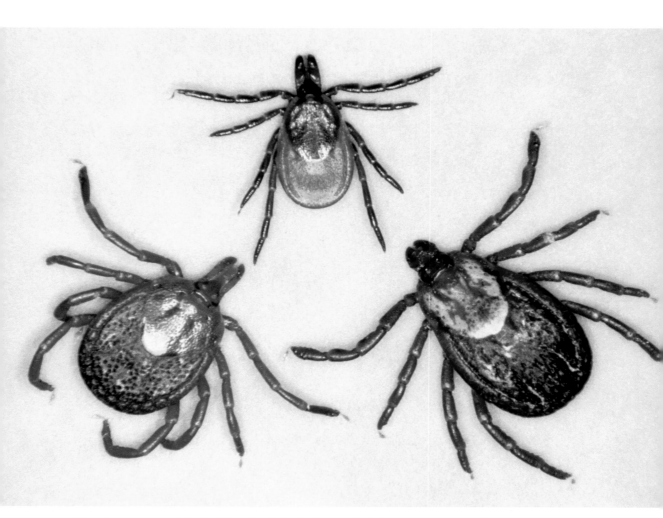

The lone star tick (left), *the black-legged tick* (center), *and the American dog tick* (right) *are common in the United States.*

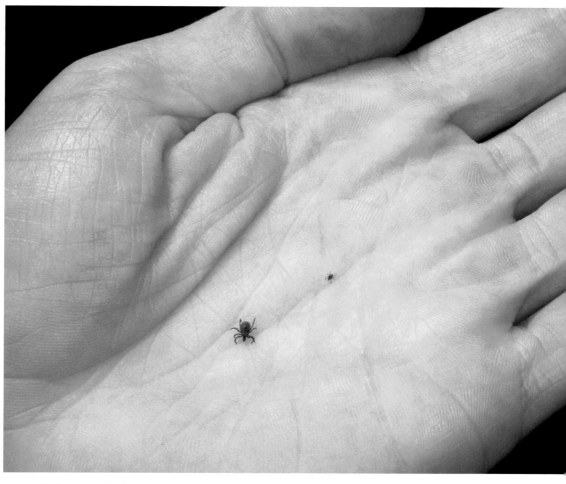

An adult tick (left) *is much bigger than a baby tick* (right).

Ticks come in many sizes. Big ticks can be 1 inch long. Small ticks are about the size of the period at the end of this sentence. Many ticks are about the size of a pencil eraser.

Female lone star ticks have a white spot on their backs.

A tick's body may be red, blue, brown, silver, black, or other colors. Many ticks are more than one color. Some tick species have special markings on their back.

Most tick species have a hard plate called a scutum (SKOO-tuhm). They are called hard ticks. The scutum supports and protects a hard tick's body. The scutum covers part or all of the tick's back. A hard tick's mouthparts stick out beyond the scutum.

The red and white area on these hard ticks is their scutum. It covers only part of the female tick's back (left). But it covers all of the male tick's back (right).

Some tick species have no scutum. They are called soft ticks. Wrinkly skin covers a soft tick's body and mouthparts.

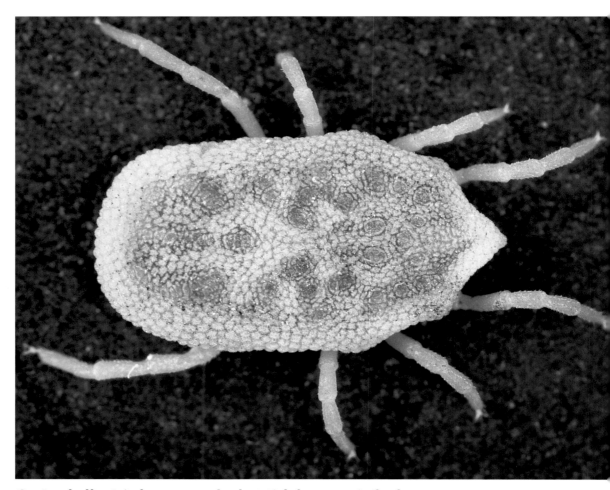

Carios kelleyi *is known as the bat tick because it feeds on bats. The bat tick is a kind of soft tick. Soft ticks belong to the Argasidae family of ticks.*

This tick is using its legs and claws to stay on a piece of grass.

Ticks have jointed legs with claws at the end. Ticks use their claws for climbing and holding. A tick uses its legs to crawl slowly. It usually does not crawl more than a few feet from where it was born. It also uses its front legs to sense what is nearby.

Hard ticks and soft ticks do not have heads. But their mouthparts look like a head. The mouthparts are called the capitulum (kuh-PIH-chuh-luhm). *Capitulum* means "false head." Ticks use their mouthparts for holding onto and feeding on animals.

A tick's capitulum is small and looks like a head. It includes a tick's mouthparts.

All ticks are parasites (PAIR-uh-seyets). Parasites live and feed on other animals. Ticks attach themselves to other animals and drink their blood. Ticks need blood to live and grow.

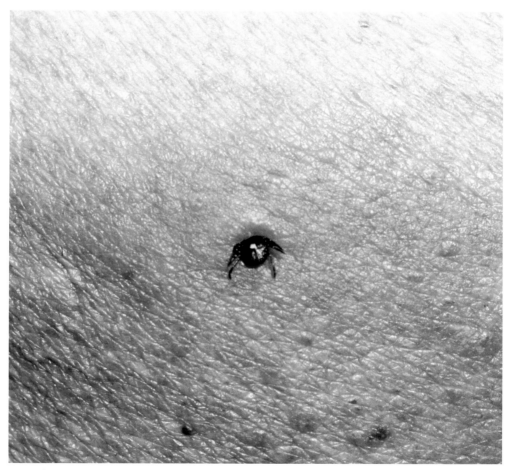

A tick sticks its mouthparts into another animal's skin to feed on its blood.

Chapter 2

Ticks have lived on Earth for millions of years. They might have fed on the blood of dinosaurs. Where can you find modern-day ticks?

Where Ticks Live

You can find ticks almost anywhere in the world. Most live in grassy and wooded areas. Ticks often stay in piles of dead grass or rotting leaves. It is warm and damp there.

Shrubs and bushes are also favorite places for ticks. But ticks cannot stay there too long. Their bodies would dry out, and they would die.

Ticks often hide in places with tall grass.

All ticks spend some time living on a host. A tick's host is a creature with blood. Birds, deer, mice, and cattle make good hosts. Ticks also feed on lizards, horses, people, and any other animal with blood.

Dogs, cats, and other outdoor pets are often hosts for ticks.

This tick is crawling on a blade of grass.

A tick may climb on a blade of grass or low branch until a host comes by. The tick may wait in piles of dead leaves. It waits and waits for a host. Some ticks wait months or years for a host.

A tick finds a host by sensing it. Host animals breathe in and out. Breathing produces a gas called carbon dioxide. When the host comes near, the tick senses the carbon dioxide through its front legs. Its legs also sense body heat, movement, and odors.

A tick uses its front legs to find a host.

From its blade of grass or hiding place, the tick waves its front legs and waits. When a tick waves its front legs, it is sensing whatever is near. This process is called questing (KWEH-stihng). The tick grabs any skin, fur, or feathers that brush against its legs.

This tick is questing. It senses a host and waves its legs.

Once the tick gets on an animal, it crawls around. It chooses a spot to eat. Then the tick has a new place to live. It has a host. The tick travels wherever its host does.

A tick uses its legs and claws to crawl around on its new host.

A white-tailed jackrabbit sits outside its underground nest. Soft ticks wait in the nest for the jackrabbit to return.

Some soft ticks do not quest. They live in the nests, caves, or underground homes of other animals, such as mice or skunks. The ticks never leave. When the host returns home, the ticks crawl on it for dinner.

Chapter 3

A tick has three types of mouthparts for eating. How does a tick use its mouthparts?

How Ticks Eat

A tick has special tools for feeding. To eat, the tick first spreads apart its outer mouthpart. Another mouthpart is inside. This mouthpart has teeth. The teeth pierce and tear the host's skin. A pool of blood forms.

A tick's center mouthpart is called the hypostome (HY-puh-stohm). The tick pushes its hypostome into the host's skin. The outside of the hypostome has barbs like the ends of fishhooks. The barbs hook into the host's flesh. The barbs make it hard for the host or a hungry animal to pull out the tick.

Ticks use their claws to hold onto the host while feeding. Hard ticks also give off a sticky substance that holds them in place.

The hypostome works like a straw. The tick draws blood through the hypostome and into its gut.

A tick's spit helps keep the host's blood flowing up the hypostome. The spit also blocks pain. So the host cannot feel the tick feeding.

A tick's mouthparts cut into its host's skin. A pool of blood forms. The tick uses its hypostome to drink from the pool of blood.

An unfed tick crawls on top of a tick filled with blood.

Ticks that have not eaten are small. For example, before feeding, the brown dog tick is about ⅛ inch long. That is the width of a lollipop stick.

As the tick eats, its gut bulges like a balloon to hold the blood. After feeding, the brown dog tick is the size of a shelled peanut. That is ½ inch long.

When ticks finish eating, most drop off their host. Some will find another host. Other ticks stay with the same host for more blood dinners.

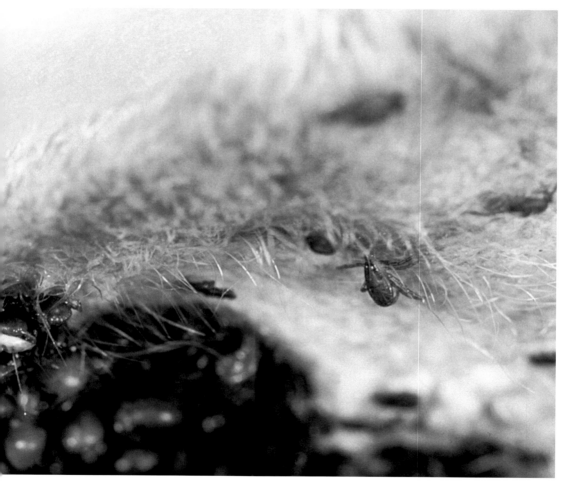

Hard ticks will feed on a host for many days or even months. These brown dog ticks are drinking blood from a dog's ear.

Chapter 4

A tick's life cycle has four stages. What is the first stage?

A Tick's Life Cycle

After eating a blood meal, a female tick lays eggs. She can lay thousands of eggs at one time. She lays her eggs in dead leaves or in plants. Once a hard tick lays her eggs, she dies. Soft ticks feed again and lay more eggs.

Baby ticks that hatch from eggs are larvas. They are also called seed ticks because they look like little seeds. Larvas are no bigger than the period at the end of this sentence. Larvas have only six legs.

These larva ticks are ready for their first meal. Most larvas choose small animals as hosts.

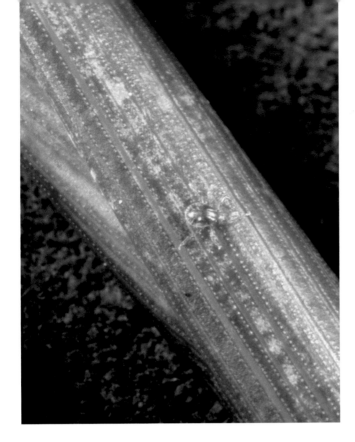

A larva waits on a blade of grass for a host. Ticks spend most of their lives waiting for hosts.

The larvas' first job is to find a host so they can have a blood meal. The tiny ticks grab onto passing animals. The larvas feed until they are too full to drink another drop. Then they drop off their host and shed their skin.

Shedding an old skin is called molting. When ticks grow, they need bigger skin. Under their old skin is new, bigger skin.

When larva ticks molt, they become nymphs (NIHMFS). Nymphs have eight legs. But the nymphs are still very small. They are not adults yet. After nymphs find a host and feed, they molt again. This time, the nymphs become adults.

A tick in the nymph stage looks like a small, light-colored adult.

This image shows a larva, a nymph, and an adult tick. Few ticks make it to the adult stage of life. Young ticks often die waiting for hosts.

Adult ticks are fully grown. They are in their fourth stage of life. Their job is to drink blood. When the females are full of blood, they lay eggs.

Some species of ticks live only one year. Other kinds of ticks can take several years to finish all four stages of life.

Chapter 5

The ichneumon wasp lays its eggs in the larvas of ticks. The baby wasps kill the larva ticks. What other animals kill or eat ticks?

Ticks and People

Ticks feed on other animals. But ticks are also food for some animals. Wasps, fire ants, and spiders kill and eat ticks. Some birds that eat insects also eat ticks. Roundworms called nematodes climb inside ticks to feed. They pass on bacteria that kill the ticks.

Humans can be enemies of ticks too. People use many methods to get rid of ticks. They cut down forests where ticks live. They mow their yards so that ticks will not hide in the grass. People also spray chemicals on lawns, fields, and forests. The chemicals kill ticks.

Cutting the grass on your lawn helps keep ticks away.

This cow is getting a special bath to keep ticks away. Cattle fever ticks carry diseases that harm cows.

People make it hard for ticks to find hosts. Killing wild animals removes another place for ticks to live. Some ticks feed on farm animals or pets instead. Farmers and pet owners do not want ticks to hurt their animals. They use medicines and bug sprays to protect the animals. Then the ticks must find hosts in other places.

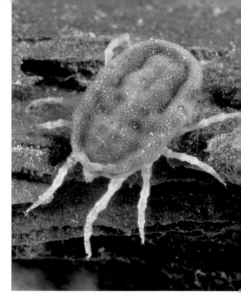

This kind of soft tick spreads a disease called relapsing fever. The tick picks up the disease when it feeds on infected chipmunks (left), *squirrels, and mice.*

Ticks are one of the biggest carriers of diseases. Tick diseases can make humans and other animals sick.

Ticks are not born carrying germs or bad bacteria. Ticks get them when they feed on the blood of an infected animal. Then the ticks pass the disease to other animals and humans when they feed on a new host.

The black-legged tick (left) picks up Lyme disease from animals such as an infected white-tailed deer.

For example, the black-legged tick lives and feeds on white-footed mice, deer, and other animals. The mice and deer carry bacteria. The bacteria do not hurt them. But when the tick feeds on us, it passes the bacteria into our body. The bacteria cause a sickness called Lyme disease.

People with Lyme disease often get a red rash shaped like a bull's-eye. The rash looks like a dot with a circle around it. People who are infected may also have a fever or feel achy and stiff.

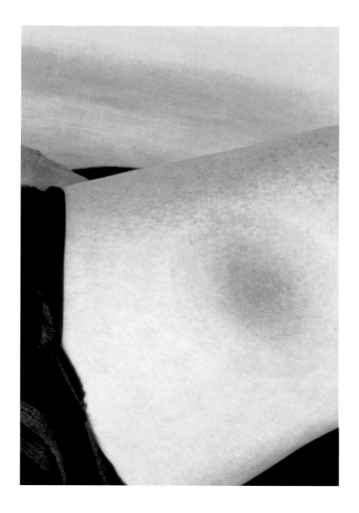

A rash like this is one of the early signs that a person has Lyme disease.

Some people try to protect themselves against ticks and the diseases they carry. They cover their arms and legs before walking or camping where ticks live. People also use bug sprays to kill ticks or keep them away.

Tuck your pant legs into your socks before walking through places with tall grass or bushes. After your walk, check your hair, clothes, and body for ticks.

These people are collecting black-legged ticks. The ticks will be tested for Lyme disease.

Scientists have been studying ticks for many years. They have learned a lot about how ticks live. They have also found ways to cure diseases caused by tick bites. Doctors are learning to spot the signs of illnesses caused by tick bites.

Animal doctors, called veterinarians, treat our pets. They give our pets medicine to prevent tick bites. They also treat other animals infected by ticks.

You can use tweezers to remove a feeding tick. Put the tweezers as close to the tick's mouthparts as you can. Do not grab the tick's blood-filled body.

Most ticks will not harm you. But you can protect yourself from ticks by wearing long-sleeve shirts and hats when playing outdoors.

Most ticks do not carry diseases or cause humans or other animals to get sick. Ticks and humans live in this world together. Woods and grasslands are wonderful places. Just be careful when visiting the natural world of ticks.

ON SHARING A BOOK

When you share a book with a child, you show that reading is important. To get the most out of the experience, read in a comfortable, quiet place. Turn off the television and limit other distractions, such as telephone calls.

Be prepared to start slowly. Take turns reading parts of this book. Stop occasionally and discuss what you're reading. Talk about the photographs. If the child begins to lose interest, stop reading. When you pick up the book again, revisit the parts you have already read.

BE A VOCABULARY DETECTIVE

The word list on page 5 contains words that are important in understanding the topic of this book. Be word detectives and search for the words as you read the book together. Talk about what the words mean and how they are used in the sentence. Do any of these words have more than one meaning? You will find the words defined in a glossary on page 46.

WHAT ABOUT QUESTIONS?

Use questions to make sure the child understands the information in this book. Here are some suggestions:

> What did this paragraph tell us? What does this picture show? What do you think we'll learn about next? Where do ticks live? Do they live in your backyard? Why or why not? Other than ticks, can you name any other living things with eight legs? What do ticks eat? What is your favorite part of this book? Why?

If the child has questions, don't hesitate to respond with questions of your own, such as What do *you* think? Why? What is it that you don't know? If the child can't remember certain facts, turn to the index.

INTRODUCING THE INDEX

The index helps readers find information without searching through the whole book. Turn to the index on page 48. Choose an entry such as *mouthparts*, and ask the child to find out how ticks use their mouthparts to eat. Repeat with as many entries as you like. Ask the child to point out the differences between an index and a glossary. (The index helps readers find information, while the glossary tells readers what words mean.)

LEARN MORE ABOUT
TICKS

BOOKS

Fredericks, Anthony D. *Bloodsucking Creatures*. New York: Franklin Watts, 2002. Learn about mosquitoes, lampreys, leeches, ticks, lice, mites, vampire bats, and other creatures that feed on blood.

Fredericks Anthony D. *On One Flower: Butterflies, Ticks and a Few More Icks*. Nevada City, CA: Dawn Publications, 2006. The author examines a community of creatures living on a goldenrod flower.

Levine, Shar, and Leslie Johnstone. *Extreme 3-D: Scary Bugs*. San Diego: Silver Dolphin Books, 2005. Microscopic and 3-D images allow readers to see ants, bees, dragonflies, ladybugs, millipedes, ticks, and other creepy-crawlies up close.

Sill, Cathryn. *About Arachnids: A Guide for Children*. Atlanta: Peachtree, 2003. A beautifully illustrated look at the lives of arachnids such as mites, ticks, tarantulas, scorpions, and daddy longlegs.

WEBSITES

Iowa State Entomology Image Gallery
http://www.ent.iastate.edu/imagegal
Click on tick names to find great pictures of different kinds of ticks.

Tick Alert
http://www.tickalert.org.au/animation.htm
This site provides moving diagrams of ticks, their body parts, and how they feed.

Tick Encounter Resource Center
http://www.tickencounter.org
This site provides information about ticks and tick diseases, as well as online games related to ticks.

Ticks: Pictures, Information, Classification and More
http://www.everythingabout.net/articles/biology/animals/
arthropods/arachnids/tick/index.shtml
This web page describes ticks with links to their arachnid relatives.

GLOSSARY

arachnids (uh-RACK-nihdz): animals with eight legs. Ticks, spiders, mites, and scorpions are arachnids.

capitulum (kuh-PIH-chuh-luhm): tick mouthparts sometimes called a false head

host: an animal whose body is used for food and shelter by ticks

hypostome (HY-puh-stohm): the mouthpart ticks use to attach themselves to another animal

larvas: baby ticks that have hatched from eggs. Larvas are also called seed ticks. They have only six legs.

Lyme disease: the most widespread disease caused by ticks in the United States. Early symptoms include a rash shaped like a bull's-eye, a headache, a fever, and tiredness.

molting: shedding old skin to make way for new, larger skin

nymphs (NIHMFS): ticks' third stage of growth. Nymphs have eight legs like adult ticks.

parasites (PAIR-uh-seyets): creatures that feed and live on other animals

questing (KWEH-stihng): the way some ticks look for a host. They climb a plant, wave their legs, and wait for an animal to come close.

scutum (SKOO-tuhm): the hard protective covering on the backs of hard ticks

INDEX

Pages listed in **bold** type refer to photographs.